Modern Farms

Activity Book

Written and Illustrated
by Jackie Nix

Written by Jackie Nix

Book Design by Karen Light of Studio Light Illustration

Publisher: Moo Maven, Prattville, AL 36066.
Visit us at www.moomaven.com for
resources and further inspiration.

You can contact the author at JackieN@moomaven.com.

Printed in the United States of America

ISBN: 978-1-7370697-2-0

This Book Belongs To:

_____

I Am _____ Years Old.

My Favorite Food Is: _____

My Favorite Animal Is: _____

My Favorite Color Is: _____

Pig Facts
Color-N-Learn

A momma pig is called a sow.

A baby pig is called a piglet

Sows usually have 6-12 piglets at a time

A group of piglets born at the same time is called a litter.

corn

Soybeans

Strawberries

Hamburger

Hay

Lemons

Eggs

Cotton

Almonds

Pie

Food Match

Cows have a special kind of stomach called a rumen that allows them to eat things that people cannot.

Can you match the foods correctly?

Which foods are good for cows and which are good for people? Which are good for both?

Cut the food squares to the left and place them in the circles below.

Good for Cows

Good for People

Good for both

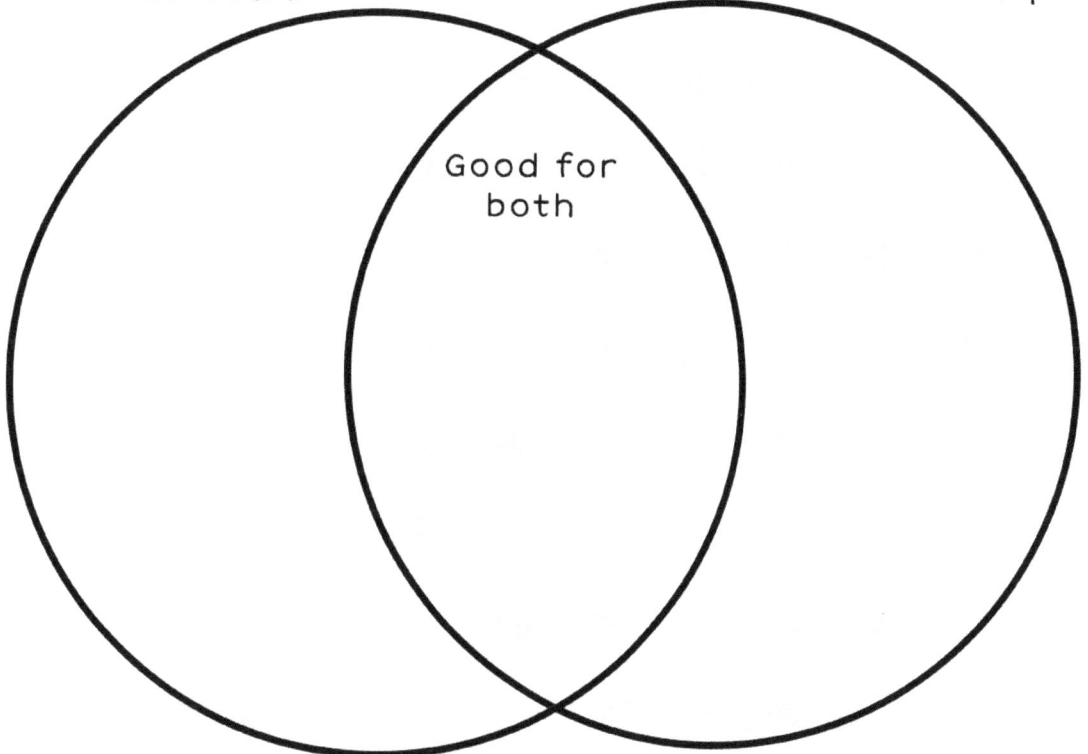

Food Match Key

Good for Cows
Hay - Cows eat hay* and grass* and get nutrition from them but people can't.

Good for Both
Lemons - People eat lemon juice and fruit while cows eat the citrus pulp* left over.
Soybeans - People eat edamame and tofu and cows eat soybean meal.
Corn - Both people and cows eat corn meal, corn oil and whole corn.
Almonds - People eat the nuts while cows can eat almond hulls* and ground shells*.
Cotton - People eat cottonseed oil (popular on fried snacks) and cows eat cottonseeds* and hulls*.

*Microbes in a cow's special stomach called a rumen, digest the fiber in all of these feedstuffs. Then the cow digests the microbes. Because people don't have these same microbes we can't eat the same things.

Good for People
Strawberries - The high water content of strawberries make them impractical for cows.
Eggs - Cattle are plant-eaters and don't eat eggs.
Hamburger - Cattle are plant-eaters and don't eat meat.
Pie - While a cow could probably eat a little pie, it isn't really good for them. It probably isn't good for you to eat too much pie either!

How did you do?   Were any of these a surprise to you?

Good for Cows   Good for People

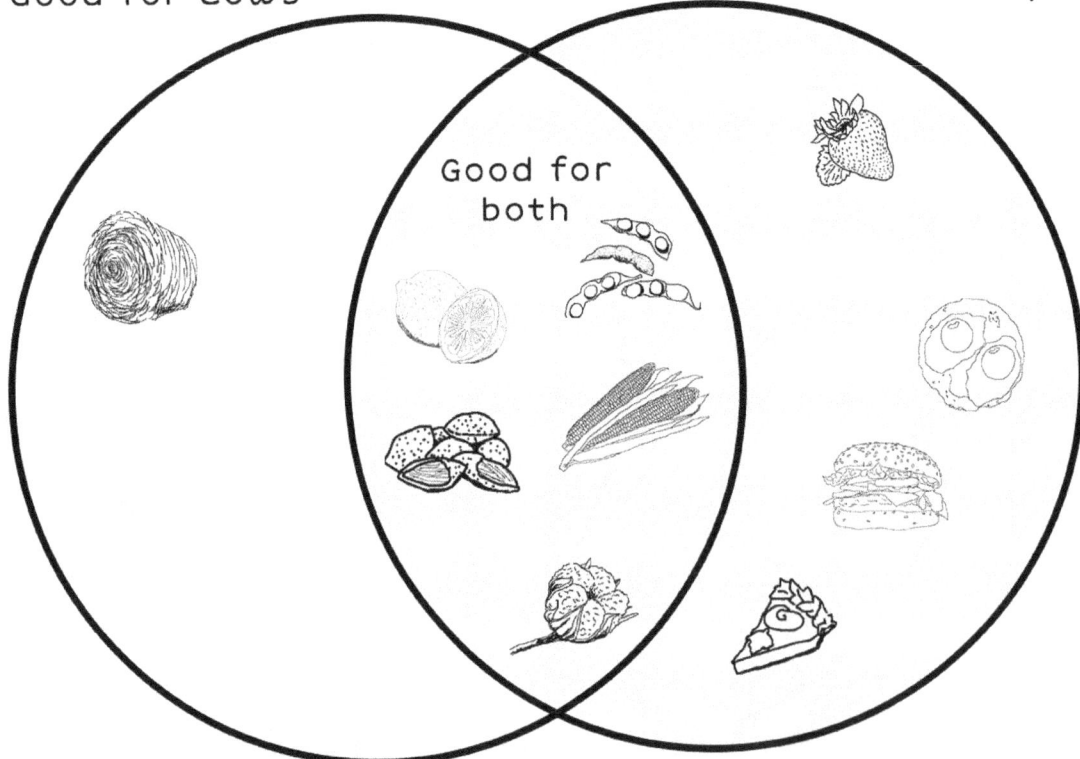

Good for both

Food Counting Fun

Count and color the correct number of food items.

1

2

3

4

5

CONNECT THE DOTS

Connect the dots by counting I through 40 to find the farm animal that provides all of these items that we use in our everyday lives.

Steak

Football

hamburger

ball glove

candle

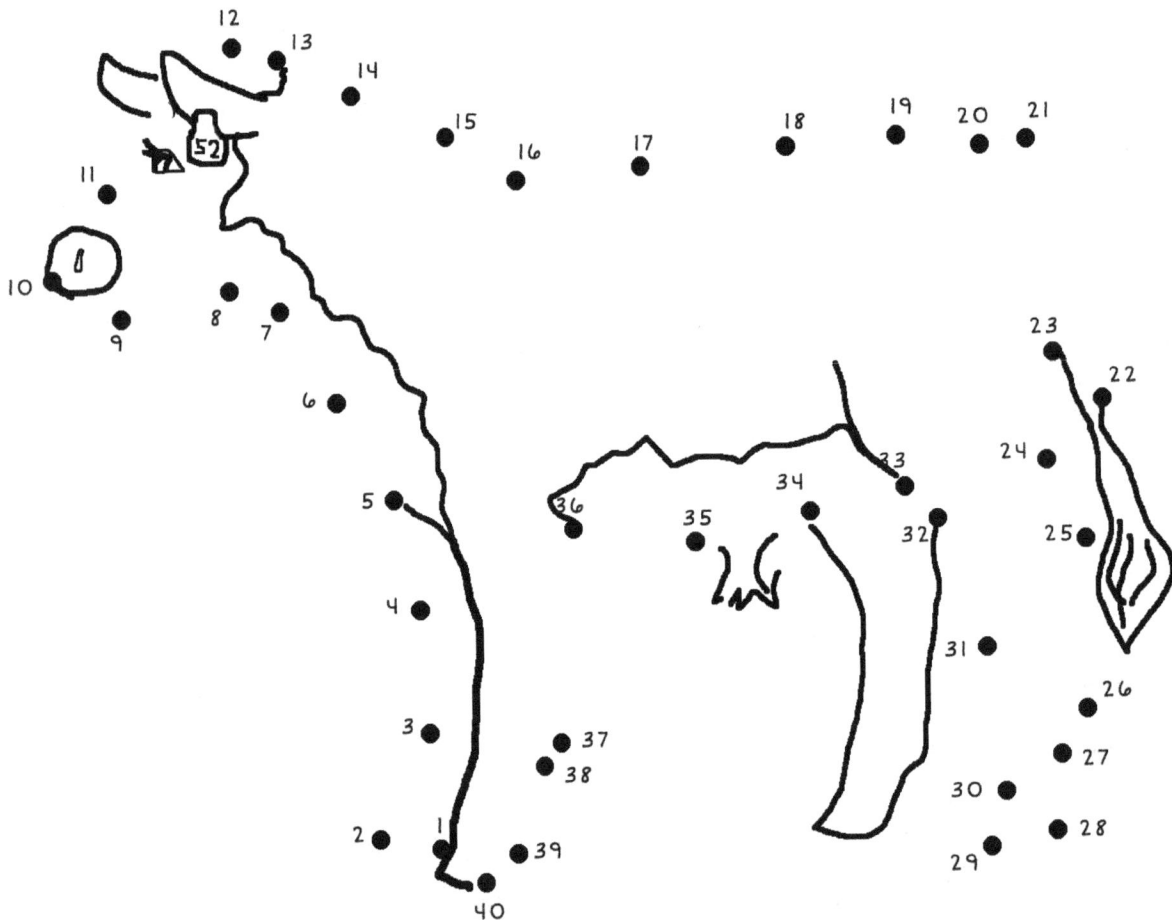

What am I? _____

Farm Animal Word Match

Directions: Draw a line from the picture to the
correct word describing the picture.

Catfish

Pig

Turkey

Sheep

Cow

Chicken

Catfish Facts Color-n-Learn

Eggs

Fry

Fingerling

Adult

A catfish lay from 2,000 to 100,000 eggs when they spawn.

Eggs hatch in 5 to 8 days.

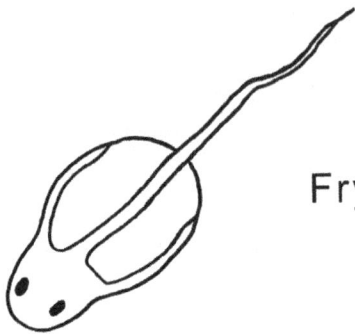

A fry is a baby fish just hatched from an egg.

A fingerling is a young fish usually the size of a human finger.

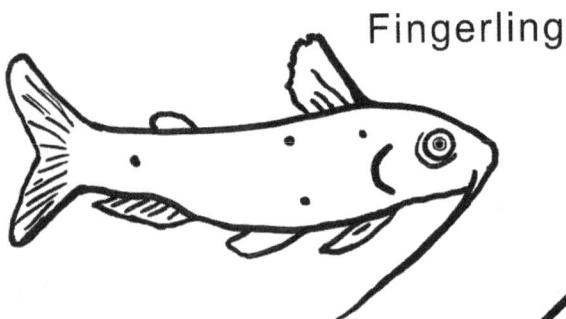

Adult catfish must be 2-3 years old before it can spawn.

HOW
Does Milk Get To ME?

Color-N-Learn

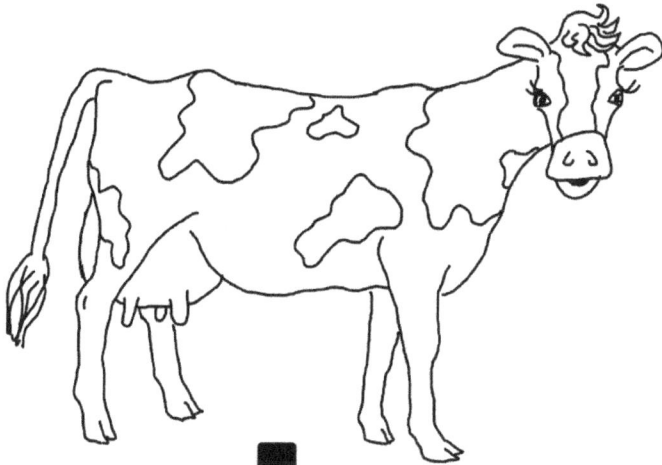

5 to 8 minutes to milk a cow

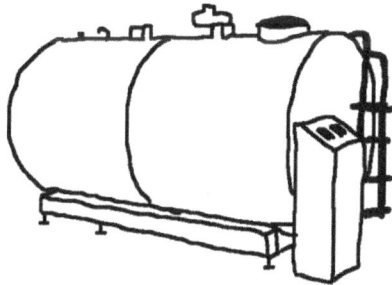

milk is immediately cooled in a bulk tank on farm

refrigerated milk trucks pick up milk from the farm every 24 to 48 hours

MILK

MILK PLANT

milk is usually processed within 24 hours

GROCERY STORE

SALE

OPEN

SALE

Most dairy cows are milked 2 to 3 times per day.

The average U.S. dairy cow makes 6 to 7 gallons of milk per day.

Milk travels from cow to the grocery store in an average of 2 days.

What is HARVEST?
Color-N-Learn

Definition
noun - the time of year when crops are ready to be gathered.
verb - to gather a crop.

We typically refer to autumn as harvest time because this is when a large portion of crops are ready to be picked. However, in the U.S. we have such diverse crops and growing zones that something is being harvested almost every month of the year!

Color these crops that are harvested during autumn.

Pumpkin

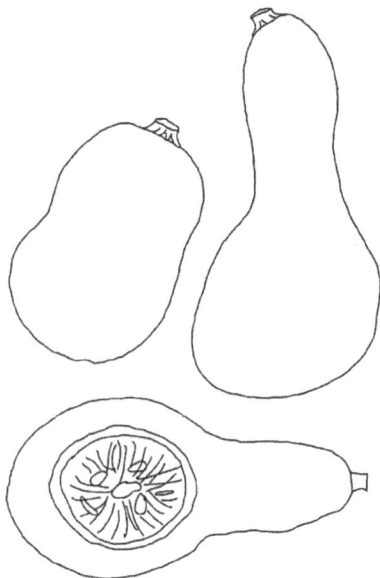

Apple

Squash

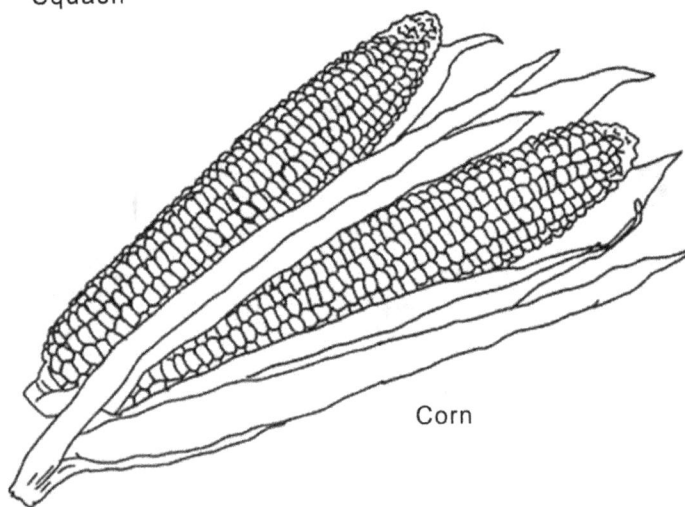

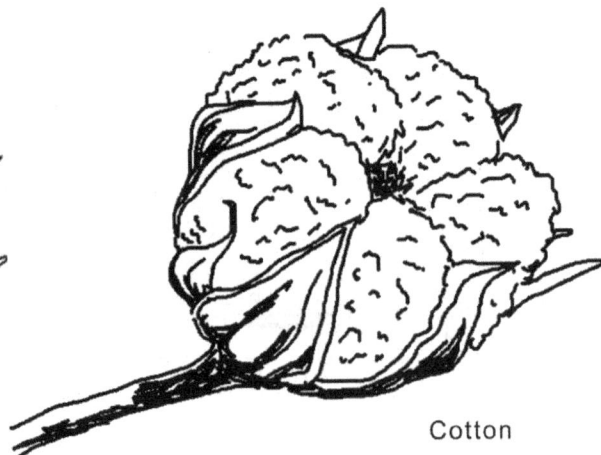

Corn

Cotton

Maze Fun

Help the bull get to the shade tree on the other side of the field by finding your way through the maze.

start

Make a Word

D A C I T W P O G

Use the letters above to complete the words below.

☐ ☐ ☐

☐ ☐ ☐

☐ ☐ ☐

☐ ☐ ☐

Find-A-Word

Find all of the words in the list. Words may be horizontal, vertical and diagonal.

Beans	Milk	Egg	Dog	Apple	Tractor	Cat
Pig	Pork	Beef	Lamb	Weeds	Cow	Goat
Calf	Plant	Cat	Grapes	Potato	Harvest	Farm
Hog	Hoe	Food	Chicken	Egg	Turkey	Bread

```
D A C I T W P I G B W
O P J A M F A T X E E
G P O Q I O C R V A E
E L F R L O O A K N D
B E G G K D W C T S S
P L O Z R P O T A T O
L C A L F A H O E H F
A X T M U D P R B O A
N Y O U B B E E F G R
T H A R V E S T S E M
P A C H I C K E N L R
T U R K E Y B R E A D
```

Story Time

Write a story about your favorite food and the type of farm where your favorite food comes from and then draw a picture about it below.

_____

_____

_____

_____

_____

Food Matching Fun

Draw a line from the food item to the plant or animal from which it originates.

Apple Juice

Hamburger

Chicken Nuggets

French Fries

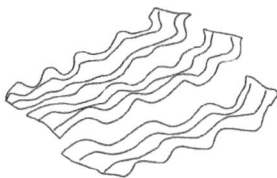
Bacon

Pig

Potatoes

Steer

Chicken

Apple

Color Your Favorite
Farm Animal

Farm Facts
Most farm animals are
raised on farms with
only one type of animal.

Find the Picture
That Is Different

A.

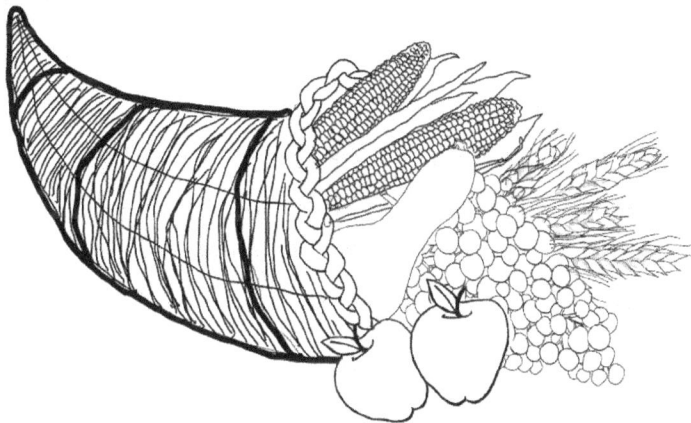

C.

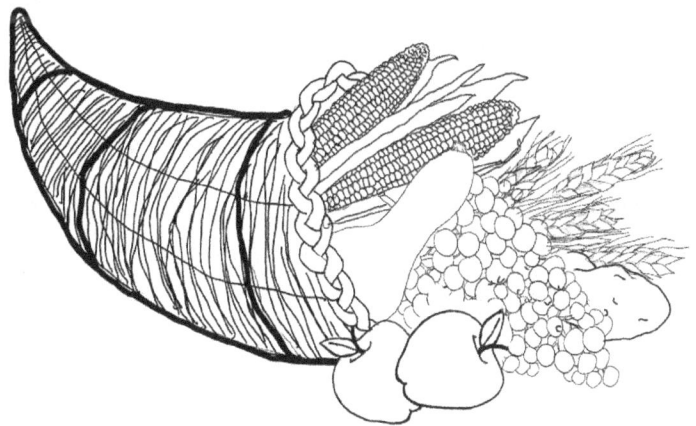

B.

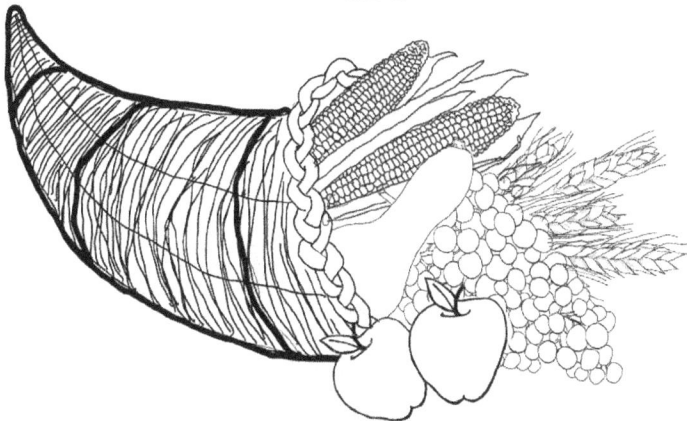

Which path leads the baby chick to its mother?

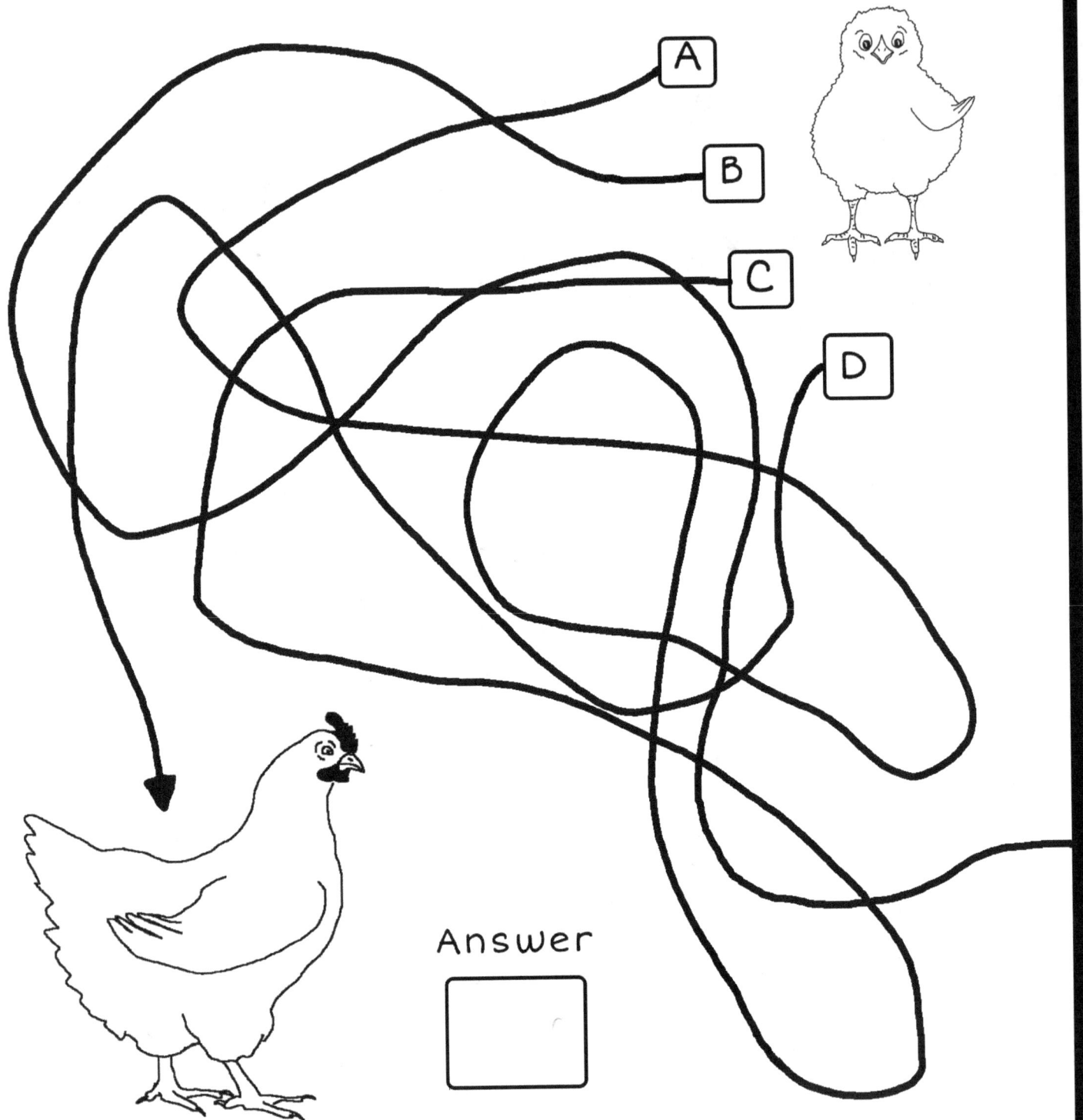

A

B

C

D

Answer

Count and Color Fun

Count the objects then fill your answer in the box.

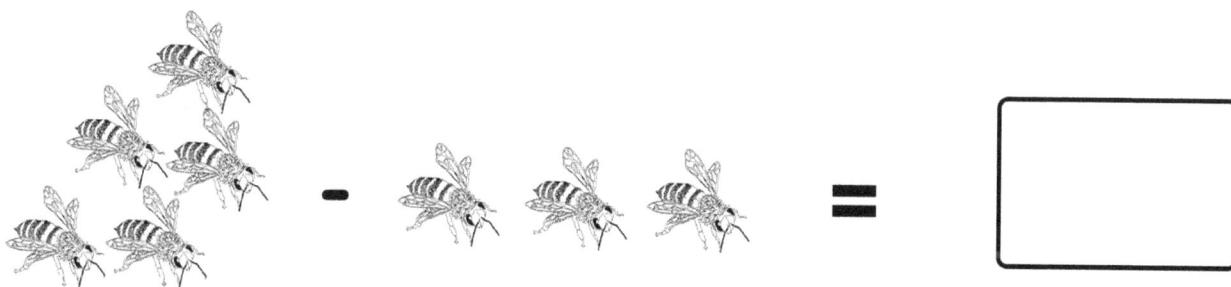

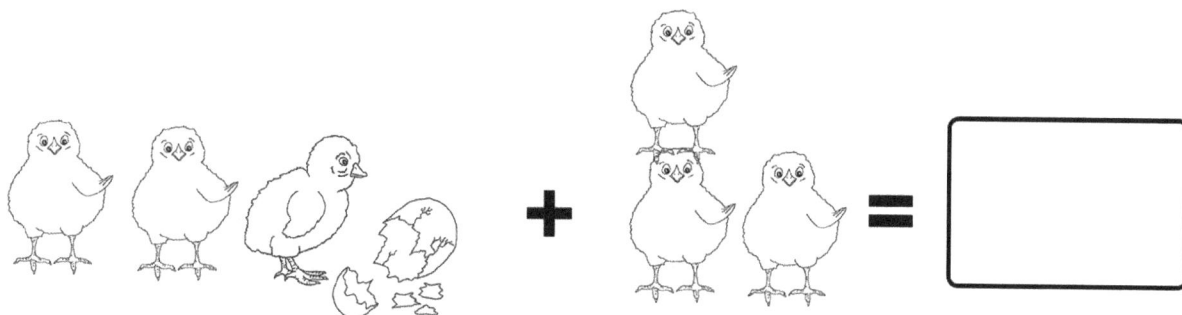

 + = [   ]

 − = [   ]

 + = [   ]

 + = [   ]

Cattle Facts Color-N-Learn

A momma is called a cow.

A daddy is called a bull.

A baby is called a calf.

Some cattle are raised for their meat and some are raised to give milk.

Some cattle have horns and some do not. Having horns or not isn't related to their gender.

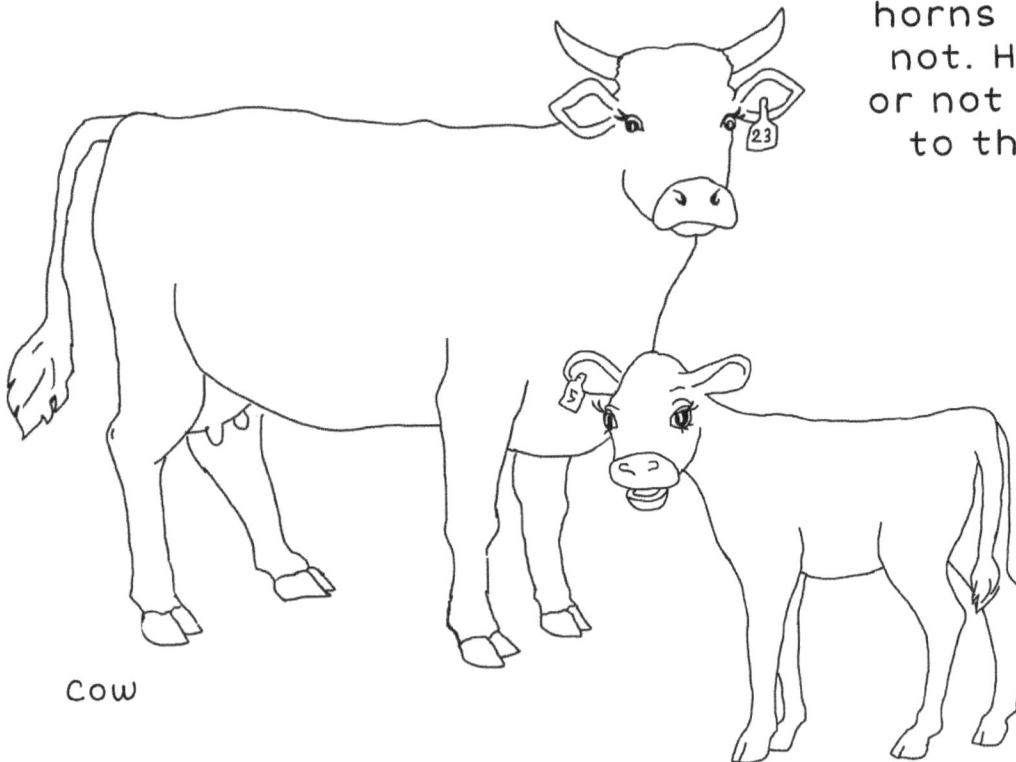

Bull

Cow

Calf

Turkey Facts Color-N-Learn

Tom

A momma is
called a hen.

A daddy is
called a tom.

A baby is called
a poult.

A group of turkeys
is called a flock.

Commercial turkeys
are raised together
in large houses.

Hen

Poult

Goat Facts Color-N-Learn

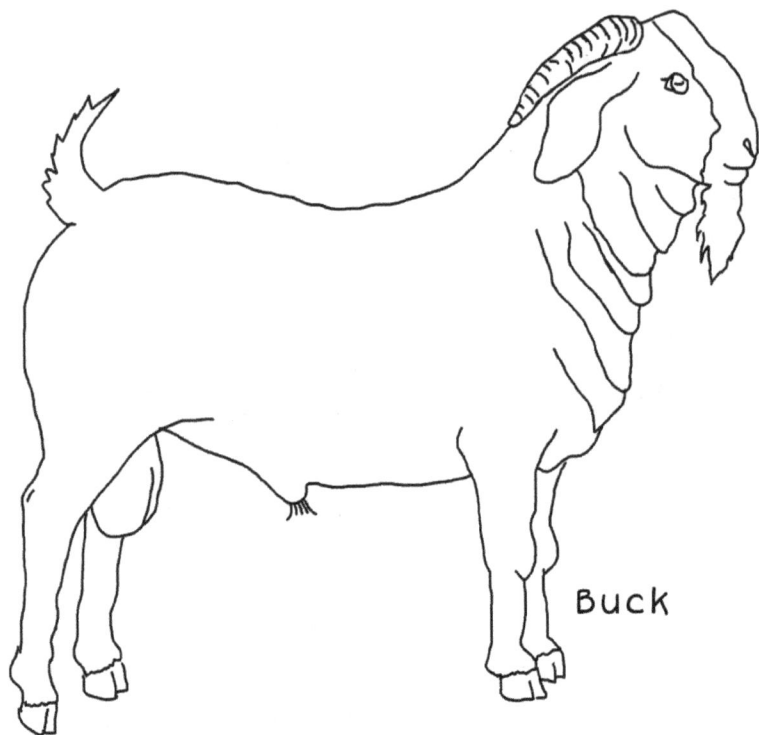

Buck

A momma is called a doe.

A daddy is called a buck.

A baby is called a kid.

Does normally have 1 to 3 kids at a time.

Mature goats weigh from 100 to 300 lbs.

Doe

Kid

Chicken Facts Color-N_Learn

A momma is
called a hen.

A daddy is
called a rooster.

A baby is called
a chick.

There are more
chickens on the
earth than people.

Chickens are raised
for meat and eggs.

Count and Color Fun

How many apples do you see? Count and fill your answer in the box.

Your Answer

Drawing Is Fun!

Can you finish this portrait of a Holstein calf?

Drawing Is Fun!

Finish drawing the scene.

Find-A-Word

Find all of the farm animal words in the list. Words may be horizontal, vertical and diagonal.

Bull	Sow	Buck	Cow	Ewe	Ram	Wether
Calf	Mare	Fry	Piglet	Mare	Heifer	Lamb
Colt	Poult	Tom	Boar	Steer	Doe	Stallion
Kid	Hen	Rooster	Chick	Filly	Foal	Fingerling

```
F I N G E R L I N G O A S
I W H X F O A L V F R Y T
L P E C O W M I E S O W A
L I N T C I B U L L O C L
Y G D Z H O N U E Z S A L
K L O A F E L A C W T L I
I E E B O A R T Y K E F O
D T H E I F E R M A R E N
S T E E R P O U L T T O M
Y O L D C H I C K V R N E
```

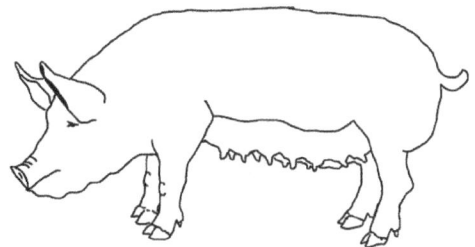

Color-N-Learn

Complete the math problems to discover the colors you should use.

7 - 5 = _____   red   5 + 2 = _____   orange

3 + 1 = _____   blue   8 - 7 = _____   green

6 - 1 = _____   brown   4 + 4 = _____   yellow

9 - 0 = _____   black   2 + 1 = _____   white